To the little ones, who encourage all of us to look into the questions we stopped asking long ago.

'Twas the night before Solstice...

Oh wait, is that right?
Yes! A party on winter's longest cold night.

Where'd they all come from, these customs of ours?
Just look up at the sky, this tale's written in stars.

Let's retell all our favorites, and follow breadcrumbs
To help us recall where these stories come from.

In the land of the North, draped in winter's stark white,
To conquer the cold, pine folk dance through the night.

They embrace the cold weather with brightest of smiles,
Season's blessings contagious for hundreds of miles.

The main reason to party? To push through the cold—
To remember the warmth of the sun's splendid gold.

Twenty-fifth of December, stars twinkle above,
Longest night of the year, spent with people so loved.

This extraordinary night marks that sun's real "birth" day,
Until June each day lengthens with more time to play.

But 'til then they must conquer this night with least sun,
So they fill it with laughter, this battle they've won.

All our timing's still set by the sun on her track,
And remind us each day and each year she comes back.

The traditions we're following leave us a trail:
Still on "Sun"days, we pause, most stop work, even mail.

And we mark the sun's axis with labels of time:
Revolutions still tallied by her bright sunshine.

On our noon and our summer, we see her bright side,
And on midnight and winter, we note when she hides.

And all over the world we see remnants of this,
Looking forward to moments when Sun's at her bliss.

In the forest, the pine folk, with baskets in hand,
Pick treats beneath pines in their own fairyland.

Standing out in the snow, holly begs to be seen,
And they celebrate nature that stays ever-green.

But their favorite of all, under big pine tree boughs,
Spotted mushroom line up, like small gifts in tight rows.

Vivid red with white spots, under Solstice tree's height,
Amanita toadstools peek through just out of sight.

All the reindeer go hunting, best scouts with soft noses,
Get excited as first spies of these winter's roses.

Mushrooms savored in winter and shared with those dear,
Always lucky to find, now known symbols of cheer.

Pine folk's baskets now brimming with nature's fun gifts,
They pack up their meal, with a wintery wish:

"May we all make it home, through the snow, safe and sound,
Then we'll sing and we'll dance and eat all that we've found!"

Through thick blankets of snow, reindeer draw their big sled,
Bright robes match their fresh mushrooms, a glorious red.

Heavy bells jingle loud, just in case they lose sight.
Rushing home to just beat the last rays of daylight.

Blending in with the land, puffs of smoke starts to show,
All their yurts draped in white, and the doors blocked by snow.

How to enter these homes? There's a secret way in,
Through the chimney they enter, a well-practiced spin.

With a pole left outside, they climb up to descend,
Friends slide down with a laugh, their most fun winter trend.

Once cleaned up from their travels, in yurts warm and snug,
They gather their goodies and give great big hugs.

The stockings are hung by the chimney with care,
Placing mushrooms inside them to dry in warm air.

They feast on through the night as their stories unwind,
Sharing all their past favorites and new ones they find.

As their grown-ups tell stories, kids drift to nap,
Near Mama in her 'kerchief and Dad in his cap.

Vivid dreams of the frolicking reindeer in snow,
Match the dancing now starting 'neath moonlight's soft glow.

All the stories flow freely, drums thud with a BOOM,
And they laugh at the antics as reindeer eat blooms.

Oh Dasher and Dancer, sweet Prancer and Vixen:
Perfect names for this spirited, moonlighted mix-in.

As they watch through the smoke they see reindeer take lift!
Celebrating the wonder of all nature's gifts.

They look up at the sky, to light years straight above
At "Pol"aris, the North Star, the orb that won't budge.

While our compasses point here, this star also shines
Perpendicular up from this North Pole in line.

All stars circling around her, we look with delight,
Just as sailors, world over, have kept her in sight.

Special shaman will follow their reindeer friend's lead
Riding off in the moonlight as all others sleep.

One by one they lie down and they drift off to dream,
Then wake up in the morning to hear what's been seen.

"What old visitors came in your dreams with the smoke?
Do please tell us what memories the mushrooms evoked!"

Generations unfolded, as customs held tight,
And we still gather 'round on this now famous night.

Though, we can't help but wonder, what thread took us here,
To the land of the cold with red mushrooms and deer?

Chuga CHOO! Brand new trains brought fresh tales to explore,
With more people to talk to than ever before.

Their loved legends now famous in lands near and far,
Shaping customs and stories, like distant bright stars.

To the wonder of all, this deep story took flight,
The original lyrics perhaps lost from sight.

We can join in together to cheer Sun's birthday
Across oceans and eons, in our own special ways.

Every year we're reminded we all can take part
In revering the past, and our loved ones, and light.

Let's find things in common, across our kind earth,
For all creatures to share; like this OG rebirth.

Greatest moral of all, 'yond enduring the snow,
Is finding the pathways to help our hearts grow.

We can always remember this gift so divine,
In the darkest of hours, this love helps souls shine.

We revel together, in love and in light:

"Happy Solstice to all, and to all a good night!"